LOST MINES
OF THE OLD WEST
BY
HOWARD D. CLARK

COACHWHIP PUBLICATIONS
LANDISVILLE, PENNSYLVANIA

Lost Mines of the Old West, by Howard D. Clark
First published 1946. Reprinted 1951 edition.
Copyright © 2012 Coachwhip Publications
No claims made on public domain material.

This facsimile reprint has no association with Ghost Town Press or Knott's Berry Farm & Ghost Town.

ISBN 1-61646-139-X
ISBN-13 978-1-61646-139-3

CoachwhipBooks.com

All Rights Reserved. No part of this publication may be reproduced, stored in a retrieval system or transmitted in any form or by any means—electronic, mechanical, photocopy, recording or any other—except for brief quotations in printed reviews, without the prior permission of the author or publisher.

Table of Contents

	Page
A Word About the Author	5
Foreword	6
The Lost Pegleg Mine	9
Pegleg the Second	20
Lost Nevada Diamond Mine	24
The Black Rock Silver Lode	25
The Lost Pete Mine	29
The Lost Dutch Oven Mine	30
The Golden Cavern of Kokoweef Mountain	33
The Lost Breyfogle Lode	38
The Lost Padre Mine	41
The Lost Cement Mine	43
The Lost Dutchman	45
The Lost Blue Bucket Placer	49
The Golden Lake	53
The "Crazy Prospector" of the Guadaloupes	55
The Lost Frenchman	57
The Lone Desert Butte	58
The Lost Clay Pipe Vein	59
The Lost Cabin Mine	60
The "Lost Nigger" Lode	62
Second "Lost Nigger" Lode	63
A King's Ransom	64
The Lee Lost Lode	66

List of Illustrations

The pen and ink sketches in this book are originals, especially prepared by the noted illustrator, Cedric W. Windas.

Pegleg the Second casts a watchful eye across the burning desert sands. He wanted no "trailers" to find his mine . . **Cover**

Exhausted, half delirious, the soldier staggered into San Bernardino with a 50-pound sack of Pegleg's fabulous ore . . **Page 21**

Through the "needles eye" hole in the canyon wall Tom Schofield saw an abandoned mining camp . . . and a Dutch oven full of gold . . . . **Page 29**

Horrified, Breyfogle escaped into the darkness, but O'Bannion and McLeod were killed . . . . **Page 39**

Linard saw the shining color of gold in the gravel beneath the lake's surface . . . . **Page 52**

One day Old Man Schippe came in and his ore bag contained chunks that were pure gold . . . . **Page 61**

HALFTONES

The Author and Ray Hetherington . . . . . . **Page 19**

Looking west across the Salton Sea . . . . . . **Page 26**

A Desert Oasis . . . . **Page 28**

The cavern shaft . . . . **Page 37**

The Author's camp on the Colorado Desert . . . . **Page 43**

A Word About the Author

Howard D. Clark is a native of Kansas and a graduate of Washburn College. Journalism called to him early in life; he has been managing editor for the Farm Press Publications of Chicago, Illinois; staff writer for a number of business papers; statistical and analytical specialist for other periodicals and concerns.

This background, plus extensive travel on the Pacific Coast, has fitted him particularly well to undertake the writing of this book. Lost mine legends make up a large section of Western folklore. In this collection he has made a sincere effort to present only the most important and best authenticated of them all.

He has also had the invaluable assistance of Ray Hetherington, an unquestioned authority in the field of Western Americana. Much of the source material used herein has been collected by Mr. Hetherington through thirty years of extensive research.

So far as is known to the publishers, this collection of lost mine legends is among the most complete and factual of any ever assembled.

LOST MINES OF THE OLD WEST, by Howard D. Clark. Published by Ghost Town Press, Knott's Berry Place, Buena Park, California. Los Angeles Office: 112 West Ninth Street, Los Angeles 15, Calif.

Foreword

It is with deep respect that the author pays tribute to the prospector. To all prospectors—those of the past especially who were apart from those of us who are "settled" in our manner of living. Here's to the old-timers! For, to an important extent, prospectors have uncovered wealth not alone in gold and silver but in metals and materials which have enhanced our ways of living, fought for us in our wars and assisted us along new paths of peace. Prospectors who receive a commensurate reward are exceptions, for the trials of the great majority will remain unsung. Here's to the old prospector, that rugged individualist, inseparable from the history of the West whether he led a burro over rippled sand or panned gravel in a crystal stream in the high woods. His modern prototype who travels with gas will do well to earn the mantle of his tradition.

The author admits license in using the term "mine" in speaking of lost mines. A gold discovery is not a mine until it has been exploited and placed upon a producing basis. It is not likely to become a lost mine then. Exceptions would be the hidden mines of the ancients, the source of Solomon's gold or the mines of the Incas and a few of the early Spanish mines in our own Southwest. But a lone prospector's diggings, however lucrative to him, scarcely rates the name which is a specific term in mining circles. Our excuse is that of convenience and popular usage. Most of us will go right on calling a dry wash a mine if it sifts a pan of flour gold. So why strip the glamour from this most romantic of occupations?

Does the reader believe he could find his way back to that quartz out-crop in a desert ledge, come sandstorm or mirage? Before passing judgment on hapless prospectors of whom there are legions, consider the ways of the desert. It is always the same, yet never twice the same. That sandy promontory, those clumps of cactus, that crumbling brown ledge may make for the same place a thousand times. But the one place that counts may not look the same again. In time, shifting sands, erosion, even earthquakes may alter the land. Vagaries of light deceive straining eyes.

The vastness of desert space is beyond the calculations of the uninitiated. Whole states could be dropped into the 35,000 square miles of desert in California alone. With two thousand of these square miles below sea level, a prospector may have only moments to sample his strike and head for water—his life in the balance. It was that way in the older days of travel on foot and in some spots always will be the same. To the novice—stubborn persistence may conquer the rock road to gold but wholesome fear is a lifesaver. Look to the water supply!

In the memory of men now living, the search for gold was a race between white men to uncover gold and the Indian to bury it, wipe out the traces and also the prospector. Their stories are of grim men, white, red, brown, of savage heat, thirst and the mockery of mirage. That the earth hides wealth, some of it known, much that is only suspected, is undisputed fact. A mere seven-line news item in the *Los Angeles Times* of July 30, 1945, tells of the discovery of a gold nugget weighing thirty-nine and a half ounces on a claim near the north end of fabled Death Valley. Breyfogle and others whose trials are told in the following stories also passed that way. The Valley itself is literally a wholesale drug and chemical storehouse of many thousands of acres containing not only gold and silver but sulphur, arsenic, soda, salt, gypsum, the borax of twenty-mule team fame, tin, gem stones and deposits of virtually every geological age.

Related, in a manner of speaking, to the neighboring Mojave Desert, these two arid tracts have been assets to the nation for many years. The Mojave has yielded lead, zinc, copper, antimony, manganese, abrasives, bentonite, diatomaceous earth and minor materials beside precious metals. Its Golden Queen gold mine long since passed the ten million dollar mark and is currently in operation. Its Atolia mine gave tungsten for the first World War. The presence of radio-active elements makes this region one of vital concern to the nation in the new atomic age now upon us.

All of our deserts have their loyal "fans" among the growing tribe of rock-hounds, amateur and professional mineralogists, and the Mojave is well up on the list. It is no place for the novice in the hotter months, yet one is tempted to reminisce—of the delicate little flowers that dot the sands in March, of the bits of old glass picked up by the writer—they were turned quite blue by the sun—of the colors of the Calico mountains, and of the desert rat's "houn' dog." He had lived out there in peace for twenty-eight years, had the desert man, with a dog for companionship. The latest member of the species to sleep before his fireplace had acquired certain social tastes in the course of the weekly visits to Barstow, which were always on a Saturday. So the dog knew when Saturday dawned on the valley and would be waiting in the seat of the car. His instinct seemed infallible (and in fact it was) but his owner disagreed on one occasion.

"You are wrong this time, pup. My calendar says tomorrow is Saturday and we will go to town then." They did, but the stores were closed. It was Sunday.

As for gold—the thread-bare adage that "gold is where you find it" is true, yet not too true. It was true only in part for the old-fashioned prospector. Many of that unschooled band knew their minerals rather well. The Golden Queen is one that came of the determined faith of one man who knew the signs and would not give up. Yet the records are heavy with tales of lost people finding gold, and then when the seekers were not lost, the gold was! In this respect history repeats itself from Texas to Oregon with picturesque variations. One of these is the fact that gold was discovered in Oregon nearly a century ago by a party in search of the Blue Bucket placer, still fresh in the minds of some members. Oregon has produced its millions but the Blue Bucket still awaits the finder, perhaps in a clear stream somewhere between Black Rock, Nevada, and central Oregon.

To him and to all seekers of lost gold the author extends this salute —more power to you!

The Lost Pegleg Mine

Long heralded as the American classic of lost gold stories, the factual tale of the Lost Pegleg is perhaps the most intriguing in all the lurid records of man's age-old quest for the yellow metal that, once seen and touched, seems forever after to elude his tremulous grasp.

A thing for dreams—this butte which was one of three. Was the whole butte of yellow-streaked black spar like the rocks so carelessly picked by Pegleg Smith—eighty per cent virgin gold? Three black buttes—were all three of a kindred breed? Some time, somehow, someone may provide the answers to these queries. For the century-old lure of the three buttes is as potent today as it was in the lusty epoch of clipper ships rounding the Horn, of hordes of bewhiskered '49ers roaring into unnamed gulches, of the days of Lincoln, Douglas, Webster, Calhoun, of a western empire peopled by Indians and buffalo.

Destiny ordered that in the stirring years of a youthful nation learning the strength that surged in its mighty rivers, that hid in the golden sinews of its roaring mountains, that waited in its broad fields, that lived in the dauntless hearts of its men and women, a story should be born. And that story would live to haunt the sons and grandsons of the men who tamed the wilderness, those who gave heed to the admonition of Horace Greely and went west. For some there are among these sons, dreamers perhaps, whose dreams vision gold such as might rival the ordered rows of well-behaved ingots now so tamely prisoned behind steel doors. May they not have their dreams—in a land where reward is for the boldest dreamer? Their sires were led across the trackless wastes by a dream too—one of a shining new world.

Call it only the natural order that sons of the fathers should feel the call to adventure and, wearying at times of the latter day humdrum, prove that blood runs true by seeking the reward of a gamble with elemental nature in a place where man is still "on his own." Thus it has been since the tempestuous Pegleg crossed the hot sands and his followers know no end.

In the interest of clarity it is necessary to explain that there were two men, each known as Pegleg Smith. Both found gold in southern California and both mines have been lost. So there are two Lost Pegleg mines, both of interest to followers of lost mine history. Mention of "the" Lost Pegleg, however, almost surely refers to the first in chronological order, the strike made and lost by the earlier of the two famous stump-legged men about 1829. His is the story which follows, told according to the best data obtainable from mining men and prospectors conversant with tradition and the picture of the land in earlier days.

Pegleg, who was Thomas Smith, was not a prospector at all until the merest chance made him one for a brief hour near the end of his career. That career, starting in his youth at the beginning of the century, would have been the answer to the prayer of any youth yearning to turn time backward to the lurid, rip-roaring panorama of the real Wild West, punctuated in the middle by the days of '49 and never to be repeated. Smith was a trader, trapper, hunter, free-booter, an adventurer who rode with Kit Carson, an organizer as well as a lone wolf when the spirit moved. His yen for living dangerously but seldom in a small way led to shooting it out in many a saloon and narrow street from the adobe villages in the southwest to the tributaries of the Missouri. The readily detachable stump leg was a weapon to defy the hardiest of assailants. His romances in town and tepee are legendary.

Dissenting against the authority of his elders even as a school boy, young Tom sounded the keynote of a colorful career when he ran away from home back in the Blue Ridge country. Working his way westward, the call of the trackless spaces came more and more clearly to his ears. Out there was the life for him, where a man could live according to his own code, without the curb of the authority he hated. Ranging from the Mexican villages of Arizona, then under the Spanish flag, to the turbulent village of San Francisco, and to St. Louis, the mecca of fur traders, the impetuous trapper and hunter fought with friendly Indian tribes against their tribal enemies, led in raiding parties against both red and white, fought at the drop of a word and spent the profits of a wilderness season in one thorough alcoholic bender at some favorite point where conduct was a personal affair. He is said to have lost his leg in an Indian battle up in the North Platte region where a companion had been killed. Smith ventured out to find the body and bury it. Shot from ambush, he fought on with a fractured leg and then amputated it with his own knife. Still a young man, the hardy hunter recovered and, carving a wooden stump for himself, strode into the part of his career that has made his name a legend wherever lost gold is mentioned.

Smith's span of life had passed most of its milestones by the time of the great gold rush. Perhaps if that exciting era had come earlier so that he would have been "gold-conscious," history might have been changed. But he was by nature a trapper and a trapper's game was a thing alive, with hide on it, not an inert mineral. However, as if to write the climax of an epic story of real life, the man who had become Pegleg, found and lost gold, his bold self-assertion in the face of opportunity failing him for the first time in any serious undertaking. Reconciled to the failure of his biggest hunt—one for which his life had given him no preparation, he died in the 1860's after a period of virtual retirement.

Historians believe it was in 1829 when Pegleg made the fateful trip from Yuma across the Colorado desert in southern California, thereby walking into the pages of history. He and an associate had led a party of trappers on the upper Colorado River and with the coming of spring and the end of the season their stock of furs—beaver, bear, deer, buffalo, was ready for the market. With Los Angeles for their destination, Pegleg and Maurice Le Duke took the load down the Colorado to Yuma for a starting place to cross the desert. Yuma was then an Indian village of undisturbed antiquity, ancient even when the Spaniards had found it nearly three hundred years before. The two men could not have left a ripple in its tranquil life to record their passing, valuable though such a clue might be.

To know their exact route from Yuma would mean riches in any land and in any language.

According to Pegleg's account they left the junction of the Gila with the Colorado to travel over the sand dunes of the American Sahara into what is now the Imperial Valley. Underestimating their need for water, they soon found their supply to be dangerously scant. The sun was already scorching the barren sands of the desert and the rocky ledges were dry of any water pockets that may have existed. After three days into the fiery waste that drops below sea level the two men began looking anxiously for water signs. They paid less attention to the general direction of travel and scanned the hummocks, they walked around the jagged rock masses seeking greener color in the wiry brush, they climbed to look across the shimmering space of heat and mirage.

At this, the most critical point in the story of Pegleg's adventure, he has been quoted variously in different interviews in his later years. He is said to have been headed northwest at the time after probably taking a westward course at the start. A far distant range of hills or mountains drew his attention. He told of climbing one of three small buttes to gain a better view of the land ahead toward the mountains. Again he said a bad sandstorm caused them to lose their way until, finding the buttes, he climbed one of them to gain his bearings and look for indications of water. His impression that he had wandered to the north is probably a reliable one. But whether he had gone north off the course or had merely come up far enough to approach it, in its more modern conception, is a question. That one has bedeviled searchers these many years. It all depends upon how far south Pegleg had been traveling on his route away from Yuma. Routes as such were virtually non-existent in 1829 and he had little more than the general direction for a guide. However, it seems clear that the two men believed their location to be three days by slow pack train to the west and northwest of the junction of the Gila when their concern over water led them to the buttes.

From some of Pegleg's remarks, the buttes could have been three humps in a low ridge, or folds in the terrain, not clearly evident except from a certain direction and likely under given conditions of slanting light. The tired man climbed the loose side of one of the black buttes and sat down on top to rest. But stones on the bare, rocky top made a rough seat. Kicking some of them out of his way, he observed their black color and that many were oddly rounded, about the size of a walnut. The top of the hill was covered with them, so much alike that he picked up one and pounded it. The black coating came off like a shell of varnish, revealing what Pegleg thought was a ball of pure copper. He had seen copper. So he put several pieces in his pocket and departed. The men had seen a high mountain to the northwest and reached it in a day of travel, finding the long-sought spring at its base. Le Duke named it Smith Mountain. After a rest the two men went on "across mountains," as they said, and reached Los Angeles with their cargo of furs.

No doubt Pegleg's report of this eventful day is as accurate as he was able to give, for he seems to have had no thought of offering misleading information. Stranger than his story is the history of the specimens and his discovery of their value. In later years he claimed to have learned this startling information on reaching Los Angeles, but it is difficult to reconcile this with his seeming unconcern about the buttes, at least for many years, for when the rocks were assayed they were found to be eighty per cent pure gold!

At the time, Pegleg sold the furs and went on his customary tall spree which, in this instance, was a bit too rough for the local citizenry. He left by urgent request and, says tradition, a substantial number of the better grade of stray horses apparently "followed" him back to his hunting grounds. Perhaps such incidents throw enough light on the character of the erratic and unpredictable Pegleg to account for conduct scarcely to be explained by a backward look. It must be remembered that he was only about thirty years old, reveling in his strength as a hunter and free agent in every sense of the word, unaware of need for money that could not be met in a season at the sport of trapping.

Also it is true that Los Angeles was a peaceful pastoral village of vineyard owners and rancheros, about one thousand in number, according to Bancroft. They were centered around the present Olvera Street, living in the golden age of the missions which have lent so much color to California history. Spain had not long since relinquished her claim to the country and all residents had taken an oath of allegiance to Mexico. At about this time the Yankee Dons were becoming evident —enterprising New Englanders for the most part. With an eye to the future they met the requirements of becoming Mexican citizens and joined the church. Most of them married lovely señoritas of wealth among the Spanish aristocracy and became first citizens in the new land, their names now designating boulevards from Hollywood to the mountains. In this atmosphere it seems clear that such an eastern tourist as Pegleg was not merely a menace to the standing of his former countrymen but a drunken rowdy deserving to be chased out of town. It was equally necessary that he spread the word that his kind was not welcome. If he had a story of gold to tell, he was not the type to have told it to the right persons, once he had the proceeds from his furs. No doubt his specimens were seen, for many perhaps truthful accounts would indicate that they and his saloon tales started treasure hunts while Pegleg went back to his trapping. But whether or not he was sober enough to grasp their significance is doubtful, even if he heard it during his visit.

During the ensuing years Pegleg found that beaver pelts were less in demand for the high hats of the effete east and for the markets of Europe; prices were falling and game becoming less plentiful. Whether or not he recalled the golden promise of the three buttes, or was prompted to have the relics assayed when the news from Sutter's Mill electrified the world twenty years later, is purely an academic question. The fact was that he did realize, at long last, that he had sat upon a golden hoard of such apparent size as to have shamed the pirates of the Spanish Main. No one was more sure than he that he had climbed upon a hill of shellacked nuggets coated as if to preserve their brightness. He had tossed them about with his own hands and climbed down the steep pile, poking his stump deep into the side to save rolling down, golden stones and all. The thirst of that day was the lesser event in his memory and he yearned to see the buttes again.

Pegleg's character and the background of his time seem to account in a reasonable manner for his attitude toward gold, especially if it is assumed that he knew the secret of the buttes. If his reactions were not true to form it is clear that he was not a prospector nor was he of the years when men sought game only to sustain the search for mineral wealth. But once awakened to the new madness that swept the land, what would he do about the three buttes? Since Le Duke had died, how would the one man who had been there go about relocating the spot? Could he apply here the resources that had taken him over the wildest aboriginal parts of the continent?

Pegleg was older now. He was losing the violent energy of the years of his towering strength. He even managed to make peace with the residents of southern California and about 1849, when every tongue spoke of gold, he made a serious attempt to reach the buttes again. Crossing the desert from Yuma to Los Angeles, he tried to retrace his trail again. But the growing tide of travel across the dunes, over the Colorado desert to Warner's Ranch, to San Bernardino, had made trails and they confused him. He could not rediscover his own.

Then he organized an expedition which would be equipped, so he thought, for a thorough search of the route and certain success. Pegleg had not been accustomed to failure. But the party included Indians and liquor which, with the white men of his choice, was a mixture to guarantee failure at the outset. Perhaps Pegleg had lost the charmed quality of leadership that had saved his life among hostile Indians and commanded the respect of the most dangerous characters of his time. While the cavalcade camped near Warner's Ranch the white men quarrelled, the Indians took some of the horses and Pegleg quit the planned trek in righteous wrath. Some accounts say he crossed the desert alone but only succeeded in reaching Yuma again. His subsequent efforts to raise a grubstake failed. With gold in the timbered waters of the Sacramento country no one would listen to Pegleg and his dry desert buttes. His fame was no asset in matters concerning money or personal responsibility. After a few years of living in Los Angeles and Yuma, even working on the sympathies of Indians who had so often fought by his side, he gave up the effort, finally taking up his abode in San Francisco, where he died in 1866.

At this point it is of interest that the Pegleg chronicle is supported by another witness whose story holds the implicit confidence of the faithful believers in the three buttes. Another man, this one a miner, found the hills of black spar and read their meaning at once. That his samples interested others is an understatement, for the limited quantity he was able to bring out of the desert in his lone trip yielded several thousand dollars in gold. But this man died before progressing to the point of action or even passing on any helpful clues except that his account of the buttes checked exactly with that given by Pegleg and the nuggets were said to have been identical. His poignantly brief moment in the picture is, however, an element in adding to the fascination of the Pegleg Lost Mine.

Taking up the search where Pegleg left it, how can one go about it to retrace that first journey from Yuma with the load of furs? Would he hold a direct route toward Carrizo in the orderly manner of travel followed by the later trailmakers, looking to the north for the high mountain to the northwest? Or might he point for the Salton Sea and, passing the boiling mud pots, visualize the thirsty Pegleg walking to the northwest over what was then a hot dry waste but is now the southwest extremity of that inland sea? The Salton Sea, two hundred and fifty feet below sea level, was formed by an overflow many years after Pegleg saw the buttes. Could he have stood upon those low, arid sands now covered with water and looked upward to where the smallest dents on the horizon appeared like hills? And from them did he see perhaps even the Santa Rosa Mountains? And then what mountains did he cross to reach Los Angeles? Or might he have gone southwest from Yuma into what is now Mexico as did Anza and his colonists in 1776, before turning to the north across the border somewhere west of Signal Mountain and south of Carrizo? Remembering that there was no U. S.-Mexican border to concern the traveler, and no established route toward the present Imperial Valley, could Anza's trail have been kept alive by Indians and migrants so as to have offered some semblance of a route to Pegleg? Since Le Duke had died before the gold era he left no record of his opinion.

Had Pegleg been holding a true direction from Yuma toward Los Angeles before he started bearing off to the north from it? And how far to the north? Aye, that's the question. It has afforded material for heated arguments lasting whole lifetimes, for this and that theory, for unfounded tales of what Pegleg said, for dark hints from Indians, for fruitless searches and wanderings by amateur and professional prospectors, some of whom have spent their lives in the quest. Superstition has been called upon for aid. Men have walked blindfolded, hoping to get lost as Pegleg did. Others have followed the whims of the prospector's best friend, his burro, and few will have the heart to find fault. Many are the campfire tales of that patient, intelligent and mayhap stubborn and arbitrary little beast. If boasts can be believed its long ears can turn deaf to threats and supplications of its owner while it leads him on a chase that like as not may end beside an outcrop of gold-bearing quartz. True it is that the burro has led thirst-stricken man miles across trackless sands to a water-hole, though not from wholly disinterested motives.

Supplementing his first explanation of how he found the three buttes, Pegleg is said to have explained that he had planned to reach Carrizo as a main point on his way to Los Angeles. And during what should have been the last day of travel to Carrizo a sandstorm blew the two men off the trail, if trail there was, or their sense of direction. It was in this blur of flying sand that Pegleg found himself approaching the rise to one of the buttes from which he hoped to learn the lay of the land and also see a hint of water signs. The butte was one of three elevations in a low ridge formation. While he showed no doubt at any time that he was off to the north of the way to Carrizo, herein lies a mystery. His story would appear to localize the buttes to an area roughly between the Coyote Mountains to the southeast of Carrizo, Superstition Mountain to the northeast and Fish Mountain on the north with possibilities farther north.

This surmise, however, has been worked on exhaustively by many an addict of peglegitis. Whether or not the buttes were part of a ridge, a hogback or folds in a gently rising land, none of the countless humps so well prospected have yielded any varnished nuggets. Among varying opinions is one that Fish Mountain is Smith Mountain, where the men found water, but the evidence seems unconvincing. Another view with a certain appeal is that Pegleg, untroubled by a passion for accuracy, may have left the Colorado at a point north of Yuma and crossed the desert at or near the southern tip of what is now the Salton Sea, then to have wandered north into the Borrego desert, still in the general direction of Los Angeles. However, sandstorm or not, he insisted that he saw the high mountain to the northwest from the vicinity of the buttes and in view of the sudden occurrence of these storms, this could well be true. The rate of travel in sand with a pack train could have allowed time for this event within a comparatively short distance in miles and before the view could be obscured. Further confusing the over-all picture is the fact that various gold strikes in the region that may be called Pegleg country have given rise to the claim that one or another is the Pegleg. While some have resulted in mines of record, none have shown ore of the Pegleg type or been concerned with three buttes. That the Pegleg legend blankets the terrain from the hot sub-sea wastes far north into the mountains is attested by such pranks as those of a joker, presumably from Hollywood, who "planted" a number of stump legs to worry serious-minded enthusiasts. Long-continuing evidence of gold known to Indians in the foothills of the Borrego desert supports the belief of some that if Pegleg did wander north of his planned route to Carrizo it was into that valley on the west side of the Salton and there that he saw mountains northward.

Warner's Ranch of early California history is a short distance farther west. Pegleg brought his expedition to the Ranch to make preparations for the jump-off at his chief try at relocating the buttes. The significance of this move is not to be overlooked, though Warner's was a famous stopping point for early travelers between Yuma and the coast. A time-honored home of Indians with its valleys and hot springs, wayfarers both red and white and good and bad found safety from the rigors of the eastern desert there. Later the Butterfield stage line reached it via Carrizo and the Argonauts of gold days passed on the way to the fabulous diggings in northern California, unaware of wealth hidden in the dry and barren gulches only short miles away. The Ranch was the rendezvous of desert-bound prospectors from coast settlements, its name woven into the fabric of their adventures. The annals of mining record that millions of dollars were taken from the earth in later mining operations that flourished from the nearby Julian-Banner region to points far over the Colorado desert. Inasmuch as John Warner, or Don Juan Warner, the Mexican citizen, gained his landholding of 40,000 acres from Mexico during the 1840's, the Ranch was not there to be an objective of Pegleg in 1829. But his design for his exploring expedition points to a confident belief that he had not far to go from there, perhaps to points within a day's journey from Carrizo and exhaust a circuit of that region.

Heated arguments have arisen over the geology of localized spots claimed for Pegleg sites. Evidence exists to indicate that in ancient times a portion of the sea extended from the east into the foothills of the mountains separating the Imperial Valley and the Colorado desert from the Pacific. Coral reefs and fossilized coral beds in the Carrizo Creek region, for example, have yielded specimens said to be of Atlantic coral, unrelated to any of the Pacific types. Thus at this meeting place, where mineral bearing formations jumble in wild disarray with deposits of antediluvian waters, a range of sand stone hills may look up toward hard rock. And in many lower areas the upheavals of earth crust have displaced the sea deposits with mineral lands of assorted values.

Innumerable legends of gold traded by Indians, found and held by some prospector until a saloon orgy would end his career, or lost by the soberest of men permeate the very air of this Southland. So many of these tales intertwine with the Pegleg tradition that it is small wonder that some versions of his story become colored with incidents belonging elsewhere. But above all looms the deathless story of that stump-legged man, the growing shadow of his exploit encompassing the smaller ones, the belief in his single strike a veritable religion to many a humble person, awed by the toils of the many who have tried and failed but unable to leave off. For lost gold is no respector of persons. The humblest of all may as well be the lucky finder. The legend keeps itself alive and growing through this very conviction of its truth, not from concerted action or unofficial promotion for ulterior reasons.

So the secret of the black nuggets is still guarded by hundreds of square miles of burning sands, fiery hot in summer, of cacti, rattlesnakes and mirages—the three little buttes and their high mountain to the northwest. Many of the faithful believe they await the finder out there not long away from Yuma by modern means, to the northwest or southwest or up in the Borrego country. For evidence of this belief, come fall every year and many are the "peg-legger" camps to be seen out there on the desert all the way from Mt. Signal on the Mexican border right on up around the Borrego Bad Lands and into the Santa Rosa Mountains.

Is it too much to believe that some time someone will chance upon three faintly marked humps long since so covered by the drifting sands of many another wind that only a hint to the most wary will have to suffice? And that one will excavate the golden spar!

The author (right) and Ray Hetherington, friend and collaborator, indulge their favorite hobby.

Pegleg the Second

While this story deals with a Smith of another peg leg, the gold is not "of another color." Perhaps the arm of coincidence is not too stretched for having both Smiths one-legged men, or perhaps both men Smiths. One is tempted to the observation that the name of "Smith" was convenient in such times for many a man from "furrin" parts on the other side of the mountains. Few questions were asked and fewer answered. It is enough that this was an out-and-out new Pegleg Smith, whose day in the sun came late in the 1860's, about the time of the passing of the original Pegleg and a generation after that one found and lost the golden butte.

Little is known of the background of the second Pegleg. Some miners believed he had been a '49er who sought fortune farther afield than most of that transient tribe. Not a young man, he was obviously no novice with the prospector's pick. Lucky he may have been but such luck has rarely been purely accidental. And, contrary to the oft-told story of the desert, he never lost his way back to the rich diggings where chocolate-colored quartz was studded with solid gold. Nor is there any intimation that he fell in with the habit of boasting or even talking of his strike. This Pegleg seems to have "kept still in nine languages," not counting Indian dialects.

Yet this lone wayfarer of the burning wastes was familiarly identified in settlements all the way from Yuma to Los Angeles, particularly San Bernardino, focal point for many prospectors of the time. He traded his gold for supplies, cashed it in openly with supreme confidence that no one could trace him to the source. Indeed, a thousand and one attempts at that trick met with utter failure. Pegleg had his "system." The canny desert rat would pitch camp at the driest spot on the terrain and wait. Then he would move on from one dry camp to another until bone-dry followers departed with tongues cracking. Then he would vanish into his private Eldorado and be seen only when he planked down another sack of ore that was maddening to the most hardened assayer.

The real search for the Pegleg started with its owner's passing, shortly prior to 1870. It is worthy of note that this lost mine is one of the few of record that were never lost by their owners, but were lost only to those seeking to inherit them by rediscovery. The Ben Sublett and the Lost Dutchman fall into the same category, each profitable to the owner during his lifetime.

No one knows how Pegleg died, out on the desert, another bulging

Exhausted, half delirious, the soldier staggered into San Bernardino with a fifty-pound sack of Pegleg's fabulous ore.

sackful of the familiar ore beside him. But history is clear that his remains were found by a deserter from the army post at Fort Yuma during his lonely trek to San Bernardino on foot. The haggard and delirious soldier dragged into town carrying several pounds of the well-known Pegleg ore tied in a budle. When able to talk he told the grim story of his wanderings, of thirst, hunger and mirage. He had stumbled upon the body of a peg-legged man with a sack containing about fifty pounds of gold ore beside him. His description identified the dead man as Pegleg, who was seen no more.

The convalescent soldier came under the care of a prominent physicain, Dr. DeCoursey, county medical officer and head of the County Hospital of San Bernardino County. A rather difficult patient, the soldier was slow to confide in anyone but did become friendly with a former American army trooper, an elderly man of German lineage. The two men with the doctor then hatched a scheme for a return to the scene of Pegleg's remains and a search for the chocolate quartz outcrop. But the strange fate that has beset so many seekers of lost gold struck too soon. The unruly deserter, still a sick man, ate some forbidden food he had purloined from the hospital icebox and died, pronto. Searching parties trying to retrace the soldier's trail from Yuma failed to find the remains of Pegleg or any signs of his diggings. The fine hand of the desert Indian may have appeared here, for the population turned out in force on the quest for easy money.

Apparently the soldier was unacquainted with the usual routes westward from Yuma and his untimely death prevented him from telling, or perhaps even learning, whether he had found the route to the south and west of the Salton Sea which was used by travelers enroute to San Bernardino. There is every reason to believe he did not know that route and instead, followed the old route between the Chocolate Mountains and the east side of the Salton basin. This checks also with what is known of Pegleg's activities. For while some word-of-mouth tradition would have him traversing the Santa Rosa Mountains west of the Salton to a point near the southern end of the range, these probably had origin in misleading talk on part of Pegleg himself. He was a consummate liar gifted with an air of confidential sincerity. Pegleg's knowledge of the Chocolate Mountain country, his escape dodges, evidence that Indians also knew the source of his gold there all give credance to the belief that his mine was east of the Salton, not too far southeast of Dos Palmas.

It is of interest that Pegleg came to his end on the same trail to San Bernardino that was followed by the soldier. This would support the belief that the soldier had knowledge of the old trail along the foothills of the Chocolate Mountains and past Palm Springs toward San Bernardino. This was Pegleg's route homeward from his mine, and it was apparently well toward the latter end of the journey that his body was found. The tattered soldier could scarcely have carried the heavy bundle of ore even half of that punishing distance.

Current knowledge of the one-legged man's habits indicate that he did not need to conduct mining operations in the usual sense, for his outfit was not known to include more than the tools for field prospecting. More likely he picked his gold from an exposed vein. But one important "tool" he had was water—unless it failed him at the last.

Specimens of the Pegleg ore still in existence not only verify the main story but indicate that it probably rates well up among the richest known to the annals of mining. Some were more than half gold, with dark brown quartz suggesting an iron-impregnated deposit. It might have been a stray outcrop that would play out on production. Who can tell? The lumps, from the size of a walnut to half a man's hand, were not placer. The extent of a vein of such richness would be something to dream about! But first where is it? Many have died trying to solve that one.

Lost Nevada Diamond Mine

Supposed location: About twenty miles south of Las Vegas, near the Colorado River.

Time: 1872.

Contrary to the usual run of lost fortune stories, this one deals, not with a luckless prospector, but with a successful miner, too successful to worry about identifying a pocketful of unfamiliar, shiny rocks. The gold and silver of tradition were his stock in trade and he knew his ore when he saw it, particularly silver, which he found in abundance.

In 1872 this prospector, named Lawrence, arrived at the old Stewart ranch, now the site of Las Vegas, on a prospecting mission. Heading toward the Colorado at a point where it would be about twenty-five miles away, he came upon an odd blue streak in the volcanic formation. As nearly as he could judge, much later, it was about twenty miles from the Stewart ranch. At the time he took only a few pounds of the blue mud from the seam which ran at right angles to the surrounding formation. Later he washed a small pocketfull of "rocks" from the mud and went on with his search for silver, one which resulted in a successful mining operation.

After carrying the batch of crystals for some time the thought occurred to him to ask a jeweler if the rocks were worth anything. They were. All of them were first-water diamonds, one to three and one-half carats in weight. Not in need of funds and in great faith that he could find all the diamonds he wanted, the affluent Mr. Lawrence gave away the stones to his friends. But imagine his embarrassment at never finding the blue mud seam again!

Perhaps the outcrop had been covered by shifting sands, the trails obliterated, new ones made, even the contours of the landscape altered. Such is the too-familiar ending of many a dream of wealth in the uncharted desert country. Yet, shifting sands shift again and uncover as well as cover.

The Black Rock Silver Lode

It was about 1849 when the wagon train of emigrants arrived at Black Rock, in northwest Nevada, after following a much-used trail. Fremont's mention of this prominent point had made it a landmark easily found by the pioneers and the hot springs in the region assured feed for their livestock. According to members of the party who settled in California, a strange adventure befell them here, giving rise to one of the oddest tales in the long saga of silver discoveries.

While the outfit rested at a spring the meat hunters went into the nearest mountain in search of game, ascending to a plateau from whence they could look down toward camp. Later they lost their way and in coming down a slope leading, as they thought, to camp, they encountered a great deposit of volcanic ash. Floundering through the loose ash, they found slabs of pure silver, large and heavy, some too big to lift. Apparently volcanic fires had melted the ore and this natural smelting process had left solid masses of pure metal scattered over a considerable area. The hunters picked up more of the heavy lumps than they could carry and were compelled to drop some of them on failing to find the camp where they expected it.

One of the hunters was a well-known character known as Hardin. In the following years, while Indians made life too uncertain for less than a sizeable army, his story got around and became familiar to many. Later, other emigrants came upon some of the silver lumps left by the wayside, but gained no knowledge of their source. The slabs carried to the trail's end were said to have been familiar objects in the settlement, known to jewelers and craftsmen interested in working up the metal into useful items. A number of years later a searching party returned in force but failed to find any traces of the silver or even the ash deposit. Other futile expeditions from the original caravan and from succeeding wagon trians found the landscape changed during the years by landslides, avalanches and storms. Generations of prospectors have come and gone, but if there was silver, the secret vaults of the mountains looking down on Quinn River desert have again locked it away.

Viewing the considerable foundation for the story of the silver slabs, one might wonder why such hardy men as could cross the wilds to California would not soon venture back for such a marvelous bonanza. But they had grown tired of looking death in the face, at least for some time to come. Indian warfare was at its height and tribes of the region were especially aggressive, roaming abroad in overwhelming numbers.

Survivors of the long trails and night watches knew the odds against any party that could be mustered. Forty members of a wagon train were killed in a battle in 1850 while attempting to cross Black Rock Desert. A short distance to the east the army officer in command of the forces of the territory was killed from ambush. Peter Lassen, of trail-blazing fame, and a companion were killed near Black Rock. Battle Lakes to the south were named by Walker, who won an engagement by killing thirty-nine Indians. This fight took place near the spot where the Paiutes demanded tribute for the murder of one of their chieftains. They got it—with thirteen hundred Indians to emphasize the request. According to history, they received one thousand dollars and a wagonload of supplies.

So, most of the emigrants had found enough of "promised land" to wait cheerfully for times to become safer before hunting lost treasure in the wilderness. The women folks, remembering lonely graves by the trail, war whoops at dawn, wanted no more of it.

Looking west across the Salton Sea towards the Santa Rosa Mountains. This fabled and beautiful area is linked with many of the lost mine stories of the West.

The Lost Dutch Oven Mine

Some foundation exists to support the argument that the long-sought Dutch Oven digging has been found and is being developed into a producer. However, this is difficult to verify in face of the fact that the second and last man to lose it had not, at last report, agreed on this point. If it can be proved that this storied lode has been rescued from the limbo of fable, then one of the more imaginative of gold stories has come true, to refute the argument that lost mines belong in the same category with the "fish that got away."

When Tom Schofield was a young man in the 1890's he worked at the task of maintaining a water supply for the Santa Fe railway company. His adventures in gold started while he was acting in this capacity at Danby, in eastern San Bernardino County, California. His story has been familiar for nearly half a century, of how he explored the folds and ridges of the Clipper Mountains immediately to the north and one time finding traces of an old trail. The trail led to a water source, then dimly among hills and into a canyon where it apparently came to a dead end. Tom's search then revealed a "needle's-eye" hole between two rocks or through the canyon wall into the site of a deserted camp. Here were mining tools, scattered remnants of the equipment of an established camp, showing every evidence of long neglect. Tom saw this location only once and under the pressure of an understandable excitement, but he has not spoken of finding bodies, bones or evidence of a tragedy at the spot to account for its desertion. But only an untold event elsewhere could account for the pile of rich gold ore in plain sight. A trail led to a shaft on the hillside and more ore.

Among the items of the deserted camp was an iron Dutch Oven— and the oven was full of gold! Tom took all the gold he could carry and found his way back to Danby. Later he and a partner outfitted for a trip back to the site, but the needle's eye gateway was not to be found. For fifty years, man and boy, Tom has looked for that hole in the rock. So have a horde of searchers from far and wide, numbering into the thousands, during the years the story has been in circulation.

Just who is supposed to have found the elusive entrance is not clear, but some mining men insist that the site has been verified in the main details of camp, shaft and ore. But with the shocking difference, to Tom Schofield, that the claimants tell of finding it or a remarkable facsimile in the Old Woman mountains south of Danby instead of in the Clippers on the north. Did Tom look in the wrong mountains all that time? It seems a bit unlikely and Tom isn't saying. Ironically, U. S. Highway 66 and the Santa Fe railway pass through Danby between the two small ranges affording the traveling public a view of both.

Has the Dutch Oven digging actually been found? Likely as it might appear at first glance, any concern over the denouement of the mystery of the old bake oven has been manifested chiefly by outsiders. The steps leading to the present stage of the reported development have been the routine transactions between small claim holders and the mining company which leased their claims. Schofield made few comments. After all, the area is known to be rich in mineral values.

A desert oasis. Spots like this saved the life of many a prospector in olden days —still come in handy on many occasions.

Through the "needle's eye" hole in the canyon wall Tom Schofield saw an abandoned mining camp . . . and a Dutch oven full of gold.

The Lost Pete Mine

The golden ball of the sun had just sunk behind a jagged hill and the quick dusk of the desert was forming shadows which leaped into being behind rocks and scrub brush and then merged into the dimming light of evening.

We three prospectors, part-timers in 1942, lighted our campfire and got out the beans and bacon. We were in a hurry to get this eating business over with and go to work with the ultra-violet lamp in a night-long search for tungsten. Time was short, tungsten was an essential material and the "blacklight" of the lamp makes it glow in the dark, so the cooler hours of the night gave us our best chance for finding it.

Our camp spot had been favored not only by us but by many who knew of the protected site at the old Red Raven Mine in the Shadow Mountains west of Adelanto, not far from Victorville, California. The old cabins there had seen much use, so we were not surprised when a voice hailed us and its owner walked into our circle of light. An elderly, grey-haired man he was, but with the poise and easy directness of years outdoors.

"Call me Hard Luck Weidemeier," he explained, as greetings passed around the fire. "Been through here before and figured to make it for camp again tonight."

So we invited him to put in with his grub and as we ate, the talk ran to mining and prospector's luck, good and bad, both of which had been known to our visitor. Of course we got around to talk of gold, the inevitable topic and ultimate hope of every prospector worth his salt, and the conversation warmed up over the last of the coffee. Soon it developed that Hard Luck had hit for the Klondike at the first news of that great strike in the "days of 'way back when." It also happens that our senior member can boast of his youthful adventures on the Yukon and is a veteran of the famous Chilkoot Pass, the Waterloo of many a tenderfoot. So the reminiscences began to fly and out of these came a story that left us little time to prospect for tungsten that night.

"I knew a fellow up here," said Hard Luck, as he settled down with his pipe. "We was chummy and hung together when we went in to spend our dust. It was pretty wild, those days. I called him 'Herb' and that's all the name he used, and I didn't think much about such things then anyway. From Arizona he was, and once when he was sick he told me one I'll never forget. Fact is, that's why I'm on my way down there now. I've lived upstate a long time and once before I went to

Arizona for this, but hadn't any luck. Gettin' old now, but still want to take one more look around, and it will be the last one. Herb got well and went off to a new strike right away, but if he'd ever found this Pete's mine in Arizona I'd have heard of it some way.

"Herb said this Pete was supposed to be a Frenchman. Seems like several Frenchmen come along and found gold and then got lost or killed and left a lost Frenchy gold mine in Arizona at one time and another. Herb nearly died looking for this Pete's mine himself and his two partners did die. I've told his yarn a lot of times and I know a lot of fellows that went there, but nobody's found that gold yet. For that matter, I told it to Shorty Harris one time when I met him in 1907 and he was half interested, but with what was happenin' in the Race Track country he didn't want to leave there and that was the end of it. I figgered that with his luck and his knowledge anything could happen. I reckon you'll either find gold or you won't find it, no matter if you kill yourself tryin'. Don't mind if I tell this yarn again, because unless I find it this time I'll spend the rest of my prospectin' days back upstate, where we have water and don't go loco.

"Pete's lost mine is somewhere down in the Gila Mountains southeast of Yuma, in Arizona. Herb had hunted it off and on a long time by himself and then he struck up a partnership with a couple of fellows. They'd nosed around on the same hunt, too, and each had his own ideas like everybody has that's ever looked for it. They put in with their outfits and decided to prospect a strip. The strip was too big, but all three of 'em had their own notions. Their plan wasn't so bad though, and they divided up the strip into three pieces and each of 'em took one. They was to ride a strip apiece and each man circle his, lookin' for signs, and narrow it down and at sundown he'd make a signal smoke. You've prospected in the desert and know that's plenty important with heat and uncertainties and of course they hoped one of 'em would find the signs.

"They started ridin' down their circles all right, but one night soon there was only two signal smokes instead of three. That meant some kind of trouble and the two men had to meet and go hunt for the one that didn't smoke up his signal. They didn't find him that night, and I don't know how long it took, but there was a lot of hard ridin' in that lava and rock and they was pretty much beat and their horses too when they picked up his fresh trail. Gold was scattered along and it was Pete's gold. Of course they got all excited over the gold, but it looked bad for the man, and when they got to him he had gone plumb loco. His horse was dead there and the man was tossin' gold around that he'd filled a pants leg with before he was loco. There was a skeleton right by a ledge there and some old dried-up saddlebags full of the same kind of gold too. This was what they'd looked for all the time and they knew the Pete mine would be somewhere around. But the sick boy couldn't tell anything, so they loaded him on a horse and started back so as to save him and get the story.

"That night at camp the crazy loco got hold of a gun and shot one of the two that was tryin' to save him and the other shot back, so they both died. Herb was the only one left. He went back and tried to locate the Pete mine right away, but couldn't find anything and so sick of alkali water he like to never made it back to a settlement, and afoot at that. He made another try or two and got sick again and the docs told him to work somewhere else for quite a spell or he'd be another pile of bones out there. That's why he was in the Nome rush. I'd like to know what's ever become of him."

And so our visitor finished the telling of Klondike Herb's tale at our camp in the Mojave Desert. But the audience of three did not do much prospecting for tungsten that night.

The Golden Cavern of Kokoweef Mountain

Think of it! "Eight miles of gold-bearing black sand for an average width of three hundred and fifty feet and an average depth of eight feet along the banks of a river." That is the sworn statement of the finder of this legendary mine. And the river of the golden sands is in a great cavern two thousand feet underground!

The discoverer's affidavit also tells of an assay of the sand by a prominent chemist of Los Angeles showing the gold value of $2,145.47 a cubic yard at the old gold price of $20.67 an ounce. But sit tight while you read this story. Don't pack the old kit bag for a mad rush to this bonanza. It is in the hands of a syndicate and no shares are for sale. Nor will this once-lost mine be lost again.

At the time of this writing no one can say that the tale is true in actual detail, but reasons exist for suspecting that it has at least some foundation in fact. And operations are under way to learn the truth. If even a small part of this fantastic yarn is verified it would seem to belie conviction of the tired wage-earner that the last chapter of gold, of new unseen gold, has been written in musty books, and to deny his plaint that days of opportunity are over. And this story of how an Indian legend came to life will have its value, for the yarn is more far-fetched than the wildest of lost mine stories cooked up by a campfire. And those who are ready to laugh off all legends of lost lodes and vanished placer beds may think again of the great cavern beneath the floor of the desert.

Imagine finding this stream of cool water flowing over its golden sands down in vast, dark depths of a secret cavern—yes, beneath the Mojave Desert, where many a thirst-crazed gold-seeker has left his bones to bleach under the merciless sun! Imagine it, a scarce sixty miles south of the searing bottom of Death Valley! Imagine finding it almost alongside the concrete ribbon of U. S. Highway 91 (and 466) only sixty-five miles from Las Vegas. The artisans of Boulder Dam, twenty minutes away by air, would not have dreamed of it. Remember, too, the old days of the Spanish Trail, a few miles away, beaten again by the feet of Fremont and his men in 1844, and the Garces route of 1776, not twenty miles to the south, followed again in 1827 by Jedediah Smith and his crew. And give a thought then to the real pioneers of the desert's mineral wealth—those grim, lone, slouch-hatted men of pick and pan and patient burro—the misty horde of prospectors who passed this way through the long years and vanished into the sunset.

Perhaps no one will ever know just how long the cavern of the Ivanpah Mountains, near the eastern border of California, and of Kokoweef Peak, in particular, was known to the Indians. However, its known history begins years ago when the man whose affidavit, given below, was a small boy. Two Indians known to him and his father on their Colorado ranch gave the boy a map to hidden treasure, so they said, telling him that when he grew up he could get rich. Coming as it did, this news was given no more credulity than one might expect. An Indian legend that might have something in it, some time. The boy, now known as E. P. Dorr, kept the map, "grew up" and followed its directions. There was something in it. Dorr claims that the story told by the Indians appeared to be true.

Originally there had been three Indians, brothers, and from tribal history they had found again the small entrance to a cave so vast that no one could know its extent. Far down inside, down, down, was a river of rushing water. Along the banks was much gold, the Indians said. It was mixed with the black sands so no one could say how much was there. The three brothers had sifted and carried away much placer gold, but the tragedy of the yellow metal finally reached down into the cave and struck. Once while carrying their primitive torchlights, one brother fell down over a great cliff in the darkness and his brains were dashed out on the rocks below. According to tribal tradition, the two remaining brothers were forever barred from returning to the scene of the death, hence their gift of the map and its story to young Dorr.

Dorr's story is told in the following statement, sworn in an affidavit on November 16, 1934, and published in the *California Mining Journal* of November, 1940. The statement is reprinted through the courtesy of that publication.

DORR'S AFFIDAVIT

"This is to certify that there are located in San Bernardino County, California, certain caverns. These caverns are about 250 miles from Los Angeles, California. Traveling over state highways by automobile the caverns can be reached in a few hours.

"Accompanied by a mining engineer, I visited the caverns in the month of May, 1927. We entered them and spent four days exploring them for a distance of between eight and nine miles. We carried with us altimeters and pedometers to measure the distance we traveled and had an instrument to take measurements of distance by triangulation, together with such instruments . . . to make examinations, observations and estimations.

"Our examinations revealed the following facts:

"1. From the mouth of the cavern we descended about 2,000 feet. There we found a canyon which, on our altimeter, measured about 3,000 to 3,500 feet deep. We found the caverns to be divided into many chambers, filled and embellished with the usual stalagmites and stalactites, besides many grotesque and fantastic wonders that make the caverns one of the marvels of the world.

"2. On the floor of the canyon there is a flowing river which . . . we estimated to be about 300 feet wide and with considerable depth. The river rises and falls with the tides of the sea, at high tide being about 300 feet wide and at low tide about 10 feet wide and 4 feet deep.

"3. When the tide is out there is exposed on both sides of the river from 100 to 150 feet of black beach sand which is very rich in gold values. The sands are from 4 to 11 feet deep. This means there are about 300 to 350 feet of rich bearing placer sand which average 8 feet in depth. We explored the canyon sands a distance of more than 8 miles, finding little variation in the depth and width of the sands.

"4. I am a practical miner of many years experience and I own valuable mining properties nearby which I am willing to pledge and put up as security to guarantee that the statements herein made are true.

"5. My purpose of exploring the caverns was to study the mineralogy in order to ascertain the mineral possibilities and actualities of the caves, making such examination in person with my engineer to determine by expert examination the character and quantity of mineral values.

"6. I carried out about 10 pounds of the black sand and 'panned it,' receiving more than $7.00 in gold. I sold it to a gold buyer who allowed me at the rate of $18.00 per ounce. Two and one-half pounds of this black sand I sent to John Herman, assayer, whose assay certificates show a value of $2,145.47 per yard, with gold at $20.67 per ounce.

"7. From engineering measurements and observations we made I estimated that it would require a tunnel about 350 feet long to penetrate to the caverns, one thousand feet or more below the present entrance, which is some three miles distant from my property.

"8. I make no estimate of even the approximate tonnage of the black sand, but some estimate of the cubical contents may be made for more than eight miles and the minimum depth is never less than three feet. They are of varying depth—what their maximum depth may be we do not know."

Needless to say, publication of the above affidavit caused a flurry in mining circles at the time, but nothing came of it. Obviously any development of such a property called for extensive resources. Dorr's story changed in minor details from that told in the affidavit but still stretches credulity. The two men told of climbing 1,200 feet down from the opening on top of Kokoweef Mountain between overlaying limestone and metamorphic rock beneath it. The cavern below was of unknown size and the small stream, after a few miles, finally plunged over a precipice 3,000 feet high, or that much farther down into the earth!

Dorr and his partner filled pockets with specimens of sand, but the steep climb was too much for the partner, so Dorr had to assist him and was reviving the man when other prospectors appeared. Some of the sand was spilled and likewise the secret. Dorr climbed down and set off dynamite charges at two points to seal the cave, the upper one 300 feet below the opening. That level is as far as anyone has been able to descend and the engineers decided against reopening the natural entrance. Instead they have started to drill at a lower level for the sake of a shorter shaft. At the time of this writing no one has peered into the strange depths below.

Meanwhile the traditional fatality of gold has struck once more. Dorr's partner died, from natural causes. And Dorr is out of the picture. The two men spoke of rushing winds in the cave. They believed that the draught could exist only because of a second opening which they imagined they saw, far away in the roof, apparently offering easier access. Both were so convinced of this that they allowed their registered claim to the original entrance to go by default. Another prospector then staked the claim and sold it. Meanwhile Dorr has failed to find a clue leading to another entrance.

Fantastic as this story sounds, both Dorr and his partner have told it separately to persons unknown to each other and in substantially the same version. Inquiry has disclosed that even the Indian brothers of the tale had bank accounts of proportions. And a desert river out there does disappear from the surface. The mining firm in control at the time this is written is already operating a producing zinc mine on the same mountain, almost right at the spot. In a talk with a member of this company, he told the writer—"we don't know what we will find. We just discount the whole thing one hundred per cent and then we know we at least will have a lot of fun."

Discussing the Aladdin Cave angles of what is a venture into matter-of-fact hard rock with an outlay of modern cash, the writer said: "When you get the gold out, just let me have the tourist rights." For if there is a 3,000-foot waterfall under Kokoweef Mountain, then an intriguing bit of scenery would seem to have escaped from Yosemite Valley and hid underground. Yet, the geologists who have walked over the terrain say there can be little doubt of the existence of a giant cavern down there—how big is only a guess. Could it be another Carlsbad? They are definitely interested.

If a half of the Arabian Nights elements of this sworn tale materialize, the most modern of realists will concede a bit of sympathy for the lure that drew the swashbuckling conquistadores over endless reaches of desert and dale in search of the Seven Cities of Cibola.

Down 200 feet at the point where entrance is said to have been blown shut.

The Lost Breyfogle Lode

Supposed location: Western foothills of the Funeral Mountains, Death Valley, California.
Time: 1862.

News of a silver strike not far from Virginia City, Nevada, traveled fast even in 1862. The word came down into California, where it was heard by three men as far away as Los Angeles. These men were "all ears" for such good news, so they started at once. Many conflicting versions of their unlucky expedition have been told, but old-timers from the dry reaches of the Mojave hold to one they say was told them by men who talked with the lone survivor.

The three, Breyfogle, McLeod and O'Bannion, supposedly made their jump-off from civilization at the friendly San Fernando Mission. Traveling light, due to lack of worldly goods to impede them or even supplies to sustain them, they took the shortest and most dangerous route across mountain and desert northward in the direction of the new silver bonanza. It took them over the wastes of the Mojave and somehow they crossed the Panamints, whose slopes gave them a view of the fiery furnace that is Death Valley in summer. But at a camp near a waterhole Indians crept up and killed two of the party, leaving only Breyfogle, because he slept a distance away from the others.

The terrified man ran for life, shoes in his hands, right down into the Valley, and escaped the attackers in the darkness. Crossing to the east, he is said to have found alkali water and filled his shoes with it, limping his barefoot way on into the Funeral Mountains. There the story of the heat-crazed man becomes hazy, and unfortunately, this stage of his adventure has the most to do with his finding and losing riches. He turned toward what he thought was a sign of water and on the way observed bits of rock float containing gold. Not far away was the vein itself, plain even to his departing senses. Of course he took samples, all according to the last instinct of a prospector. Then, not finding water, suffering from his alkali drink and the miles on bare feet, his conscious mind deserted him for the most part. Yet his dimmed faculties took him in a northerly direction until a rancher met the dazed apparition of a man still carrying the precious ore, days later. Accounts of his fearsome condition have echoed down the years. Only an "iron" man could have lived through Death Valley in midsummer, barefooted, foodless, with only alkali water.

Breyfogle headed a party to retrace his steps, but Indians headed

Horrified, Breyfogle escaped into the darkness, but O'Bannion and McLeod were killed.

off that attempt. Later, after a change of season, he led another group to where he had eaten mesquite beans and, as he thought, to where he had seen the sign of water. But signs had changed and perhaps the first was only a delusion. Search was abandoned in favor of the known sure thing of silver back home. Perhaps a cloudburst had washed down over the slopes of the Funeral Mountains and covered the vein. Prominent mining men have placed more than ordinary credence in Breyfogle's story, supported as it was by physical evidence of samples from a rich quartz outcrop. But none believe that known mines of the vicinity of more recent date can qualify for the lost vein, so it remains among the fairly accredited lost bonanzas.

The Lost Padre Mine

Far-famed among traditions of lost mines is the story of the Lost Padre. According to prevailing belief, this is the one mine among numerous so-called padre diggings that properly owns the name, since it is the one located by an early padre and is supposed to have been worked extensively by Spaniards and Indians in earlier days. The authentic Padre vein is believed to be the one rediscovered in the 1870's by Dr. B. F. Bragg, having characteristics mentioned in old reports. The Padre was said to yield a chocolate red quartz ore rich in gold. Has it been found? Positive confirmation would hinge upon old Spanish records but an interesting development re-opens the question.

In 1941 a Mrs. Annie Rose Briggs headed an organization for the purpose of developing a gold strike which was confidently reported as the Padre. Claims of the members covered an eleven-foot vein eight miles from Neenach, California, at Sawhill Mountain. The vein at least answers to the description of it given by Dr. Bragg. Contrary to the usual practice of concealing all evidence of an important find and taking a secretive attitude, Mrs. Briggs voiced the opinion at that time that nearby areas held mineral possibilities and even invited prospectors to help develop the field.

Not one but many legends of precious metal are linked with the journeyings of the padres—early Spanish priests who were also explorers. They were pioneers of the western wilds, often the first to come with knowledge of the worth of gold and the art of refining. It is not strange that tales of lost padre diggings crop up in widely separated areas. These were not so much lost to the priests of a century and a half ago as that findings were impracticable at the time and commercial exploitation was not a purpose. Explorer-priests recorded their travels in reports to church superiors in Mexico and on occasion the archives have yielded leads to the rediscovery of workable mineral values.

One of these reports takes us as far north as the region of the Blue Bucket. Tradition places the source of the story in Mexico City, where an American mining man gained access to some of the historic writings. The padre had journeyed very far to the north and had been able to record the latitude and longitude of the place where he found gold, with other remarkable phenomena of nature. It appeared that his record of a certain boiling spring tallied with that reported by Fremont, who described these springs many years later. This would be in upper Nevada, not far from Pueblo Mountain. It was on this leg of his journey that the priest is said to have recorded the find of a fabulous ledge, a persistent feature in stories of padre finds. Of course the miner in question followed the route, checking many clues, but without result.

A lost padre mine in Texas was the legendary source of some of the great wealth of the Jesuits before that sect was banned by orders from Spain. Picturesque folklore claims that on their departure the mine, on the El Paso side of the river opposite Juarez, was filled with hundreds of jackloads of silver bullion and the deep shaft buried with vast mounds of dirt, moved to the spot with great difficulty. Other stories say the tunnel was filled with ingots of gold, bars of silver and relics by an early governor of the Spanish province when he had good reasons for departure. This tale has it that the entrance was filled with red soil from the river. One may be sure that any traces of alien soil in Franklin Mountain will long have been explored. Yet, lost gold sometimes demands that its truth be stranger than its fiction.

The Lost Cement Mine

Time-honored in its repetition throughout the full span of a century, the tale of the Lost Cement lode is by way of becoming a minor classic of lost gold lore. The prospector who has not heard this story and repeated it, perhaps with picturesque variations, would need to be sought in a land where yarns of treasure are told in a strange tongue. But variations or no, most old-timers of our American West have settled down to a fairly standardized, if short, version of the gold-bearing cement, so well has the legend been seasoned by age.

Several years ahead of the real gold rush to California an early wave of immigration brought three German brothers to the wild West. Trouble aplenty had dogged their weary footsteps and as they neared their goal traveling alone, it is believed that they were survivors of Indian trouble along the route. On one occasion when they stopped in a secluded mountainous nook, believed to be a cave where they took refuge for a night, high in a pass in the Sierras west of Mono Lake, they found gold. The ledge appeared to be cement, strong with lumps and wire webs of pure gold highly concentrated in the stone. Of course the three men took what they could carry and went on their way.

Perhaps there is a moral in the fact that the three men had survived all hardships of plains, mountains and lurking Indian ambush—until they found gold. At any rate, their worst troubles came after that. Two of the brothers died of hardship in the comparatively short but difficult remaining mountain travel and the third reached aid in such a condition that he had no interest in leaving the settlements again for all the gold in the cement ledge. But he gave word of the find and told one of his friends, a man named Whiteman, how to reach it. It was Whiteman's search for the lost lode that created one of California's earlier spells of gold fever—Whiteman looking for gold and the local gentry looking for Whiteman. Mining men say the cement ledge is still safely hidden, but notwithstanding the venerable quality of the report, they are not inclined to scoff at it.

An old hardrock miner who had migrated from the colder altitudes down into the Mojave country had heard reptitions of a parallel tale of gold in a cement-like formation. As he told it, Indians had appeared at trading posts not far from the same region, buying supplies with gold of similar characteristics. A white man later became the "lucky" finder of a cement ledge in a cave back in the mountains of eastern California, north of the central region. But on showing signs of a brief affluence and telling his story, the man disappeared, permanently. In those days it was one thing to find gold, but quite another to let Indians know or even suspect that one of their secret lodes had caught the eye of a white man. While no one knows what may have happened to him, he could not have conceived a better way to invite oblivion. Apparently he left no clues behind him and "dead men tell no tales."

The author's tent (circle) on the Colorado Desert. The hand of man seems puny indeed amid such majesty as this.

The Lost Dutchman Mine

Supposed location: Superstition Mountain, east of Phoenix, Arizona.
Time: Lost since 1892, at death of last survivor.

Built up by story, legend, tragedy and the imaginings of countless gold-happy treasure hunters, more of them amateurs than otherwise, the cumulative tale of the Lost Dutchman lives on as one of the better known examples of the lost gold mines in the fabulous Southwest. Just why the lure of Superstition Mountain not only persists but grows is not so clear, for the gold hidden there is matched in authenticity by other spots quite as localized. In fact, a search for a skilfully concealed outcrop in the rugged vastness of the area reminds the searcher of looking for a needle in a large hayfield. But persist and grow the story does, and its hold on imagination increases accordingly as people repeat it and believe the old frontiers to be passing.

Not to discount the age-old lure of the gold itself, possibly it is because the Dutchman's gold is seen through the haze and smoke of more than a century of bloody tragedy, over prone bodies of Spaniards, Apaches, murderers, grim-faced adventurers and foolhardy innocents, the latter heedless of the odds encountered in wild places. The aura of this background which clings tenaciously to the jagged pinnacles of Superstition is in itself a challenge to adventure; an almost irresistible magnet to the city-surfeited soul. So it is that the fifty years since the passing of the Dutchman and of Indian menace in the unfriendly canyons have witnessed a strange procession of fortune hunters. From nearly every state and even foreign countries have come those who would have the Dutchman's gold.

Prospector and tenderfoot alike have tramped the byways and tapped the rocks; with and without supplies or essential knowledge of mineral signs. As might be expected, tragedy has occasionally caught up in hunger and thirst, loss of direction, in scaling dangerous inclines. Summer heat and seasonal storms respect no one and even a bullet may warn the enthusiast that he is on his own in such fastness. Meanwhile second and third generations of ranchmen, familiar with every nook and cranny, scoff at the golden legends and stick to their business of livestock. Once an Apache stronghold, even Indians say the mountains are full of evil spirits and shun them.

Almost any number of versions of the story of the Lost Dutchman will vary in the details of its historical background. It seems that a Spaniard, one Don Miguel Peralta, who lived in Mexico during the

1840's, was responsible for the original discovery of the famous lode. When he sent a young Peralta toward the north country it may have been in search of fortune, or that he chased the young scamp away from the place, as there is a hint that his flair for romance had something to do with it. We have our choice of legends, if it matters. The young man is said to have returned with specimens of fine gold which won him forgiveness and the job of guiding an expedition back to the gold site. Other versions say that only the companion who accompanied him, or followed to see that he "kept on traveling," lived to return with the gold. But without question, gold had been found a few weeks of pack travel to the north, near a mountain shaped like a sombrero.

Miguel Peralta led an expedition back to "La Sombrera," also known later as the Weaver's Needle, found it and the vein of the rich samples. The moral of this seems to have been lost on successive searchers, not to mention the innumerable American prospectors who have mislaid a lone mountain or canyon a few days from the outfitting store. The Peraltas not only came upon their lode in the trackless wilds haunted by hostile Apaches, but sent pack trains of gold ore home to Mexico, until Peralta wealth became a matter of record in church and state.

When their adventure started in 1845 the territory belonged to Mexico, but on its transfer to the United States in 1848 the character of the enterprise changed. The gold mine was soon to be in a foreign country. The Peraltas had operated on a considerable scale, building and operating arrastras to crush ore dug from a probable eight or ten shafts in the general vicinity of the Weaver's Needle. Of necessity these were not too distant from central points where ore was reduced as much as possible for the long haul homeward by mule train. Trail markers of stones driven into cactus, Spanish symbols on cliff walls and boulders in the vicinity of Weaver's Needle testify to the comings and goings of miners. Expeditions had grown in size until the last amounted to a small army, one to end all expeditions into a land beneath another flag. Perhaps the Peraltas had gold enough and were at last mindful of the watching Apaches. But the Apaches had already decided it was one expedition too many, and moreover, the great herds of livestock could not be allowed to escape to Mexico. So the massed bands of Indians struck from all sides. According to some accounts, there were no Mexican survivors when the shooting was over. Others have it that three men escaped massacre by hiding in caves.

Then the Indians covered and buried all traces of mines and shafts except the one later worked by the Dutchman. Perhaps this one escaped notice, but the fact that they went to great labor to hide the sources of gold is evidence that the Peraltas worked a number of veins within a limited area. Learning from their object lesson, the Indians made a thorough job of removing temptation to enter their sacred grounds. In this they exercised the skill at which they far exceeded the white man even of that time, conditioned to the wilds as such men were and, in turn, more than we of this day and age. Only a major project of mountain landscaping could have caused the mining operations of hundreds of men to merge into invisibility, returning the scenes to a state of nature that matched cliffs and walled canyons so as to deceive the eye to this day. If the Dutchman's mine was hidden at all it was re-opened by natural causes or found by survivors of Peralta days through knowledge of its location. Many believe it has been covered again, and more permanently, by nature. An earthquake, soon after the Dutchman quit work, changed some of the contours in the vicinity of the Weaver's Needle. Perhaps the gods of the Apaches have decreed that no more invaders shall profit from the buried wealth of the tribal mountains. At least the Peralta diggings have been hidden by skilled hands.

The Dutchman came upon the scene in the 1880's, more than thirty years after the Peralta massacre. Actually a German prospector named Jacob Walz, this cold-blooded character was an ideal type for a real-life murder mystery. For the record, he killed upwards of a dozen men to gain and hold his treasure, one being his nephew, whose conduct failed to please the old man. Others faced his gun in ill-judged turns too close on his trail. More than likely he had heard tales in Mexico enabling him to arrive at the Hat locality, where his first victims revealed the mine by their presence. Whether or not the surprised miners were the three supposed survivors of the Peralta massacre, the Dutchman shot them at sight and took over.

From fragmentary remarks dropped by the old man in his last days the mine could have been on a northern slope of Superstition Mountain. However, the foxy multiple-murderer was tricky and secretive in the extreme. No one knows how much gold he took, but his burro-load hauls continued for nearly ten years. On retiring he told of concealing his mine with native rocks and logs to match the surrounding terrain, and so it remains. The Dutchman died in Phoenix in 1892.

The Lost Dutchman has its iconoclasts—irreverent persons who spoil a good story, who insist it was not a Peralta digging, but a new find beneath a small butte near the mountain. Some believe it has long been found in the notable mine of nearby Goldfield. But one can scarcely wish away the big-scale works of the Peraltas and the shafts the Apaches sealed so well, their general area still indicated by quaint old Spanish treasure trail markers on enduring rock canyon wall and boulder. Those cryptic drawings of snakes, triangles, sunbursts, mule-shoes, have confused more than assisted searchers. They had a meaning once. Were the Peralta about to quit because the mines were worked out? Or might the meaning still hold good—with a key to it? Many would like to know. Treasure maps purporting to tell all are almost "a dime a dozen," but no one has obtained a real key to the petroglyphs, without which their story may not be told.

The Lost Blue Bucket Placer

"In a cavern, in a canyon"—so the old song runs. The same words might be used to describe the supposed site of the Blue Bucket lost gold. Yes, but where is the canyon? Well, now, that is what prospectors, miners, amateur treasure hunters and farmers have wanted to know since about 1850. Gold nuggets were picked up under the high banks of a creek by members of an ox-train traversing the stream-bed, bound for "Oregon-or-bust."

The reality of the gold find is beyond question, owing to the number of responsible persons in this outfit, a large one of fifty or sixty wagons. While the location of this lost placer is hazy, a wholly separate source of gold never found by white men but used by Indians appears to link closely to the stories of the emigrant train and both point to a likely area. The story of the Blue Bucket cannot be omitted from any record of famous lost strikes because the tale will not "down" and gold actually was found.

In 1845 the wagon train had crossed the plains, mountains and forded streams until it was proceeding in a general direction to the northwest through northern Nevada, toward the Malheur country in southern Oregon. At the ford understood to be at the present site of Beowawe the party separated, one taking a route to California, the other continuing northwest toward Oregon. Up to this point the party had consisted of upwards of a hundred wagons, well-guarded, carrying families, equipment and heirlooms from former prosperity. The route of the Oregon train took it through Humboldt County, Nevada, an area characterized by mineralization in great variety.

According to the best understanding of members of this party the group left Black Rock and while still in Nevada, chose a route through a canyon that grew more difficult the farther they went. The way was unknown to them but seemed logical as they were aiming in the direction of the Three Sisters peaks in Oregon to use them for a landmark. Modern knowledge of what might have been a feasible route is of no avail in mapping the course of this wagon train. Little was known then of routes and passes, Indians had their say and pioneers often "got thar" the hard way.

Somewhere, not many days after passing Black Rock, in the early stages of this epic of taking wagons apart to lower them over cliffs, of jolting over boulders, some brassy-looking pebbles were seen in the creek bed of the canyon trail. The women and children were attracted by the pretty rocks—the men were concerned with growing hazards ahead. Many of the yellow nuggets were tossed into the rows of blue buckets hanging from the sides of wagons as the tired oxen inched their way over the loose gravel. As the days went on, children kept some of the "rocks" among their few playthings. Many days and miles later wagons capsized once more and blue buckets, rocks and precious articles were lost, presumably up in the Deschutes in Oregon.

Four or five years later, California gold taught the emigrants what a gold nugget looked like and the remaining playthings were identified. Expeditions back along the trail were chased out by Indians, men and horses were lost, little or no prospecting was possible for years under such trying conditions. Meanwhile the traces of a one-time trail became dim, acts of wild nature changed the character of water courses and the type of country made an intensive search extremely difficult.

A California physician had cared for injured survivors of Indian battles fought in some of the earlier expeditions and had heard their descriptions of the Blue Bucket locality. At a later time he showed samples of his own California placer gold to a visiting trapper from "over east" in the mountains. The trapper offered to take him to where "that stuff" could be picked up by the pack-load. He had wintered his horses in that canyon.

The two made the arduous trip to the spot and the doctor saw that it fitted the details told him by the emigrants. But cloudbursts had buried the bed of the channels with tons of sand, mud, brush, debris and cut new gullies. Not a trace could be found of Blue Bucket or any other gold. Perhaps the gold has long been buried from sight if new channels have been made. Time and violent elements have covered and uncovered riches over and over again, and only at opportune moments in geological history has man appeared on the scene at the right time. Perhaps some of the searchers have been misled by variations in the story which would have the gold found where a blue bucket was lost or left, far up the Deschutes. Perhaps others have sought for a canyon through which a wagon train could pass, while all evidence points to the opposite, especially after a century of erosion. The Owyhee and Malheur country has been prospected for many years by nearly every old-timer worthy of the name.

Could the following story link up with that of the Blue Bucket?

At an early day in the settlement of northern California a trading post near the Nevada line was patronized by Indians from the Nevada territory. Frequently they paid in gold dust and even nuggets of size, but answered no questions. After much persuasion one Indian was induced to take the trader back to a point near his tribe in the hope that permission would be granted for a visit to the tribal diggings. But Winnemucca, Chief of the Paiutes, sent word that another step would be fatal.

Much later an Indian boy brought nuggets to the post and, being more approachable, was persuaded to lead the two traders over the route again. But after several days of travel the lad either became frightened at the thought of violating rigid tribal law or did not trust the men. He vanished at night. The general direction of the trail led toward what is now northwest Humboldt County, Nevada, near the Oregon line. No one can guess where it ended. Prospectors have long combed the area in quest of the Blue Bucket trail, so perhaps all guesses have been wrong. Yet gold there was, on that trail to Oregon.

Lingard saw the shining color of gold in the gravel wash beneath the lake's surface.

The Golden Lake

Persistent but fragmentary reports of a "lake of gold" in California have crept into lost mine tales for nearly a century, but for the most part they have taken more of the color of fables heard by Coronado than of later-day fact. The mention of a golden lake bottom is all too reminiscent of the period of the Gran Quivera.

It remained for an aged prospector and miner to bring this tale to life, and down to reason, among others he told the writer in a spirited conversation about the rugged years which have provided comfort for his old age. The old gentleman, still owner of an outdoor voice and a once-powerful frame, was even able to substantiate his account with dim, yellowed newspaper clippings pasted in an old scrapbook.

Francis Lingard was prospecting with others in the Feather River district when one day he made an exploration tour alone. It was in the mountains between the headwaters of the Feather and Yuba rivers, north to northeast of Sacramento. High in the forested mountains he followed a stream to its outlet in a small lake hidden in a fold of the ridges. Lingard stood at the point where the stream of snow water entered the lake and saw the shining color of gold in the gravel wash beneath the surface. An important point in this story of the lost lake is the fact that the gold showing was in the crystal water of the lake itself, below the surface at the mouth of the stream. The year was 1853, an unusually dry season, and the lake was almost certain to be much below its usual level. And Lingard arrived at the latest stage of this dry condition.

At the time he gathered many clean nuggets and found his way back to camp with difficulty, not by a direct route. He was unable to locate the lake again in the short time before winter snows blocked the trails. Future searches held less and less likelihood of success as memory of mountain contour became hazy and normal water levels were restored. He was positive he could find the lake again, but even if he may have done so, its rise would have covered the gold deposit under deeper water, making the lake difficult to identify. Such lakes are not too rare in that region. By the following spring the enlarged stream could have carried down the new debris of a dry year to further conceal the showing of color at its mouth.

The disappointed man kept up the search for many years and it was toward their end that his trail passed the cabin of the youthful prospector, now past eighty, whose vigorous memory recalls many stirring events of his adventurous life.

The "Crazy Prospectors" of The Guadaloupes

Supposed location: Guadaloupe Peak, West Texas, near town of Odessa.

Time: 1880 to 1892.

"Old Ben" Sublett's strike is among the more thoroughly authenticated of the lost gold mines of the fabulous Southwest. Seemingly the old man actually gathered his bags of rich gold nuggets instead of mining them, as his age and meagre equipment precluded any notions of time and toil in the process.

Sublett was just another old-timer among the countless migrants to come west during the roaring years of the great gold fever. The toughest regions of the Rockies had felt his pick but had never yielded more than hard luck and the scantiest of provender. Privations had reduced his family to three youngsters, so the time had come for him to make his way to new fields and milder climes to the south. Perhaps the years of toil had paid dividends in knowledge of gold signs and it may also have been that he had gathered Indian lore that gave a hint of where tangible wealth would reward the disappointed prospector. For no known reasons he turned up in west Texas and located at Odessa, he and his children in rags and hungry as usual.

As soon as helpful ranchers had given him work enough to ward off starvation for the moment, Sublett turned to his old interest of prospecting again, to the amusement of the community. His crazily assembled outfit and eccentric manners earned him the title of "crazy prospector." Hard-riding cattlemen might warn him of Apaches in the Guadaloupes, but Ben would disappear into the forbidding fastnesses and return with great regularity.

One day Ben burst into the saloon and emptied a bag of big nuggets before the pop-eyed gaze of all present and from that time every man was his friend, crazy or no. But the old man was not deceived nor lacking in resources, boastful, queer and cantankerous though he was. The lean years had afforded him little reason to trust the human critter. No one was going to organize him and his mine. The skilled knowledge of the best trackers of the cow country failed them when they tried to follow the "crazy" man to his location. When the proceeds from one trip ran low he would simply drop from the sight of man and reappear soon with another bag of nuggets. His trail always disappeared into the foothills south of the Guadaloupes, almost in the shadow of Guadaloupe Peak, below the border of New Mexico—a mountain about 8,700 feet in altitude.

Toward the end of his days Ben relented once from the vengeful secretiveness he had always exhibited toward his old tormenters and consented to head a party organized to exploit his bonanza. But he became ill enroute and firmly believed he had been poisoned by members of the group. The trip was called off and the deal closed for keeps. Once he took his young son along, but the boy remembered only that his dad went down over a ledge, presumably with a rope ladder, and returned not to long afterward with a bag of gold. This would add a bit of credence to Sublett's oft-repeated declaration that he could not direct anyone to the place by word of mouth even if he tried. He was known, however, to hint at a spot a few miles to the north of Rustler Spring. His firm assertion that no one would find his gold without his personal guidance seems to have held good since his death in the '90's. But will no one ever find it, even as he did? Old Ben might have "stood in" with the Apaches as so-called crazy persons have been known to do, and gained a tip from them. At least he appeared to be immune from the common ailment of scalp-lifting, to the perpetual surprise of the countryside.

According to rumor, one or two other men were supposed to have found the Lost Ben and died amidst the customary liquid celebration, or, like Sublett's son, never found the way back. However, no evidence exists to prove that they actually found Ben's diggings. And the fact of his gold has been confirmed by people of the time and place, cattlemen, bankers, residents, so is not to be dismissed as a campfire tale.

The Lost Frenchman

Supposed location: The Eagle Tail Mountains, northeast of Yuma, Arizona.

Time: 1867.

Truly it was gold that the three tired, hungry Frenchmen swapped for supplies that day in Yuma. They needed, it seemed, grub and equipment for a considerable undertaking and one not to be entered into without ample promise of returns. It could only be for more gold. And even after all needs were satisfied their credit balance at the store amounted to a tidy sum, if rumors were true. In the main they likely were, because that balance was on hand for years, while bones supposed to be those of the Frenchmen bleached in the mountainous desert to the east.

Many were they who tried to follow the Frenchmen, for gold across the counter was no secret in Yuma or in any outfitting store of goldfield history. Some of the Mexicans were persistent, one boy especially who wangled a job as cook for the party. But evidence that he would surely mark the trail for followers lost him his precious job and the Frenchmen went on alone. The boy could report only that they disappeared into the Eagle Tail Mountains. No more was seen of them alive, but some unidentified bones found in the region have long been presumed to be theirs. Nor has the source of their gold been discovered. Yet scouting parties on the trail of Apaches did find mounds of good ore beside a trail.

Some theories link this discovery with Tenhachape Pass and it is true that most of the luckless search for the Lost Frenchman has been made through that region. But it is also true that the three men worked far afield of that area, from traces of their activities. Mining men have said that there is "gold in them thar hills," the Eagle Tails and the precise diggings of the lost men may be extremely difficult to identify. Experienced prospectors admit being deceived at first inspection of rusty looking rock which turned out to be gold-bearing quartz. Many have said that this is a dominant characteristic of ore in this region. The lode of the three Frenchmen, if one lode it was, seems to have escaped detection, perhaps by concealment of theirs or even nature's own camouflage.

The Lone Desert Butte

Told at times from the Panamints to the shade of palms of the Imperial Valley is the tale of the lonely butte and its mineral-coated nuggets. Just how such a solidly defined portion of landscape could disappear may strain the credulity of the uninitiated until they also learn the ways of the desert, which means to learn that no one is proof against its deceptions at all times. Quite understandably, some prospectors to the north favor looking for the butte in the desert near the northern end of the Chocolates, while those of faith in the mineral values to the south would have it in the desert near the southern tip of the same range. None have proved their contentions, but the southern location seems far more logical, considering certain characteristics of the ore.

The Butte was a solitary black one rising from the desert floor, its black rock supposedly containing manganese in what would be paying quantity if more accessible. A probable chimney formation may have been the source of the gold said to have been known to the Indians and found by various white prospectors, none of the latter being able to retrace the trail. Having only enough water for a foray into the region, the finders could stay only long enough to gather samples of the oxide-stained nuggets and cover the distance back to base. Then, with full equipment, the Butte could not be found.

The Lone Butte is only one of many spots to vanish from the sight of man out in the sandy wastes, a good example of desert deception. The changing direction of the light at different hours of the day will cast shadows at one time showing a contour of the earth and, at another hours, erase the spot by blending it into the flatness of miles of terrain. "Adding up" with this misdirection by light is the chance of a minor difference in the direction of the approach to a hillock, for instance, from only a slightly higher elevation but with the searcher looking away toward more distant, clearly defined hills. Lacking trails and guideposts, enormous odds are against the seeker who would come upon a location from the same point of the compass at the same hour of sun time.

Early Spaniards understood this or learned it from Indians and would establish a known point from which to watch the shadow of a projection reach a given spot, thus indicating a location or another landmark. Yet even this technique is of no avail where landscape lacks rugged features and in fact may be a monotony of billowing sand. The lone prospector, as no one else, is dependent upon his own devices.

The Lost Pipe Clay Vein

Because old man Schippe was a Dutchman, two generations of people in the King's River country have spoken of his lost lode as "The Dutchman's Gold." But so many localities have since boasted of a "lost Dutchman" mine that Schippe's is now better known for its occurrence in a clay that was shot with gold.

During the 1870's Schippe was a familiar figure, always the luckless prospector, patient, persistent, but perpetually low on funds. Few gave him credit for sufficient knowledge of mineral signs to make an important strike. But one day he came in, a changed man. His ore bag contained chunks that were pure gold, more or less coated with a clinging, fine clay which onlookers insisted was good pipe clay.

Schippe's ample bag of specimens were not placer, for the gold showed no signs of the action of water. And no other samples have ever been seen, as Schippe was murdered almost at once by thugs trying to force from him the secret of his location, and without avail. Years later one of his relatives came from Holland with clues that threw some light on the possible site, but not enough. This younger Dutchman carried on the quest until he died from disease.

According to the relative, the Dutchman had written home to Holland about his rich prospects, of how he had stumbled upon a wide clay ledge by accident. It was hidden in brush thickets high above the north side of the river, facing more to the east than the north. This was the south fork of the King's River, presumably a few days' travel up from the fork of the stream. Travel made by Schippe was obviously on foot and such a description of distance can convey only the vaguest meaning.

No one can guess at the number of hopefuls who have tried to estimate and repeat that hike during the past seventy years, but the lure remains. And the region has its advantage when contrasted with the risks of thirst and hunger known to desert prospectors. Lying between the far-famed Yosemite National Park and Sequoia National Park, the Sierras to the eastward, some of the most spectacular scenery in the world is literally in the neighborhood with hunting and fishing in the bargain.

The Lost Cabin Mine

Ask the oldest prospector among the old-timers up in the northwest for his favortie lost mine story and he is likely to come up with the Lost Cabin mine. He might also admit being one of the many seeking to inherit it, with no more luck than greeted his predecessors.

Back in the dangerous years of the frontier, three bold prospectors made for the Big Horn country, though no one knows for sure whether their "cabin" was eventually located in Wyoming or farther into the mountains in Montana. But once more, misfortune held its hand until they had struck it rich. Their gold was of fabulous quality, according to the samples of ore that came back—stained with the blood of two of the three men. Finding the lode far up a strange stream, they had built a raft to float back for equipment to work the diggings. Came then an attack by hostile Indians and the lone survivor had been too much occupied with keeping alive to give a clear account of his travels or retrace the route. The samples gave rise to a legend of the north country that has brought many a hardship to ambitious prospectors from far places, all through the years.

Another tale has its locale in the Big Horn country, but as might be expected of hardrock men, they give this one small value perhaps because it had no origin by way of a mining man. As the tale goes, an army unit was chasing the remnants of an Indian war party when a campsite gave a view toward the higher peaks of the Big Horn range. A soldier is said to have sat on a ledge of weathered quartz streaked with virgin gold. Samples were reminiscent of the Lost Cabin lode, says the legend, but hazy memories of the two localities fail to reconcile them. The few survivors of the army band were unable to retrace the route of the raid through the vast spaces of that fantastically broken, uncharted country. Some of it has not known the tread of the white man to this day.

One day Old Man Schippe came in and his ore bag contained chunks that were pure gold.

The "Lost Nigger" Lode

Two stories of Negro origin dealing with lost gold in widely separated regions are of interest, both being finds that were verified by samples. Under the circumstances of time and place, the samples could not have been other than genuine.

Found as an incident of trouble between Indians and army units of the frontier, this lost lode is named for a band of Negro soldiers whose mission it was to drive the Apaches from the Black Hills of western New Mexico. Among the undulating peaks and high levels of the west side of the range the soldiers climbed and worked their way through a great forest of quaking aspens and reached the top of a hill in order to get their bearings.

The whole top of the hill looked like gold to their bulging eyes. The quantities of samples collected by the entire company were predominantly copper, but also yielded gold in richly profitable values. But all traces of their trail had vanished by the time a return trip was permitted, and the sameness of the almost unlimited wilderness of brush caused the search to be abandoned.

In support of this story, old prospectors in southern Arizona have expressed the confident belief that the same lode was found and lost again by a white prospector named Hurste. Before returning from the hill in the great no-man's land of quaking aspens he was caught in a heavy snow storm. Hurste marked the trail carefully, but though he and his party made searches again and again as soon as possible, the storm had made countless additional trail markers of broken trees and shrubs. This jest of nature played havoc with Hurste's dream of fortune.

The Second "Lost Nigger" Lode

The second and more famous "Lost Nigger Mine" story resounds through the Big Bend border country of Texas, its legend as high, wide and fanciful as the great sparse spaces of that rock-ribbed empire. The betting odds are about even as to whether the locale is on the Texas side of the Rio Grande or over in Mexico. But the action took place on a Texas cattle ranch in Brewster County along the Rio Grande after it turns to the northeast after rounding the bend. This fantastic land is more reminiscent of something dreamed by a pseudo-science writer of far planets than one charted, even sketchily, by our own surveyors.

Down there to the northeast of the towns of Dryden and Sanderson, in the region of the canyon of Maravillas Creek, was the Reagan ranch. Hundreds of stories told from San Antonio to El Paso agree that one of the Reagan wranglers was a Negro and that the Negro told his employers of finding gold while riding the hills above the big river. The Reagans paid little attention to the Negro's story, though they saw a few pieces of the ore samples. They knew gold only when made into jewelry and so tagged. The Negro went to San Antonio, where he knew a railroad man with a thorough knowledge of minerals and mining. This man recognized the ore as a grade worth many thousands of dollars a ton. This much appears beyond dispute, and instead of taking its place as an idle tale, the Negro's report aroused the hopes of countless gold hunters and no doubt will do so for a long time.

The Reagans became interested, to put it mildly, but too late. The Negro disappeared, leaving a dozen or twenty stories of his probable fate, but few clues about the source of his gold. More than a dozen organized groups offered rewards for him and many individuals did the same, to no avail. The railroad man spent much of the remainder of his life looking for the Negro's lode, where gold-bearing rock could be "knocked off with a hammer." Arguments arose as to which side of the river should be prospected, for ranchers of the 1880's were wont to work both sides, the only question being that of bandits.

Enthusiasm about the Negro's lode continued through the early years of the 1900's and San Antonio papers published reports from time to time that one or another prospector had located it. Also, the gold that was found showed widely varying characteristics. And last, but worthy of note, some strange fatality of all this border gold appears to have reached or struck down many more after the disappearance of the Negro.

A King's Ransom

The Apaches may have known more than any other tribe of Indians about nature's secret hiding places for gold. Fate had decreed that the fantastic mazes of their weird habitat held seams of the tempting metal, of little value to them and wholly strange to the tribesmen of the plains. Warlike, more than ordinarily aggressive, implacable toward intrusion into their sacred fastnesses, the Apache instinct for self-preservation was whetted by knowledge of what most surely brought the white man —gold. A main objective then was to stop the prospector's search, kill or drive him away and conceal traces that might lead to discovery. For generations gold had meant only more hordes of hated white men.

Geronimo could not have failed to know of many secret lodes rich enough to have drawn an authentic gold rush. Tribal knowledge dating back of the advent of the Conquistadores was at his disposal. When finally taken prisoner as an outlaw he is said to have used every means at his command to effect his release, one of them being bribery. If any prisoner in this country could have offered gold for freedom it was Geronimo. Not ready-made, so to speak, like that of the Incas, but in quartz veins that still lure the hopefuls to the Apache country. Word-of-mouth tradition has placed his treasure lodes at various points in New Mexico and Arizona and perhaps rightly, but a favored theory would have an interesting source among the canyons of Yavapai County in Arizona, not far to the east of Prescott. Spaniards of much earlier days had found there something that was much less than Gran Quivera but still enough for their gold smelters in this region, like others they operated south through the Superstition Mountains to the Tumacacori country near the border. All of it was Geronimo's native heath and his final stamping ground when the wily chieftain surrendered to General Miles in 1886, over in the mountains east of Bisbee. If Indians are subject to logic, his choice, though by no means only source of gold, would be an old digging rich enough to have drawn the Spaniards, from which they were driven while in production. And Apache were skilled in covering even established mine operation, for more and more urgent reasons.

Adding interest to the theory of Apache gold in this region, soldiers posted there during the Indian wars to the east and southeast of Prescott claimed to have seen evidences of gold-bearing quartz. High-grade ore was found, so they said, alongside the bones of prospectors. Whether these dated from earlier Spanish exploration or our own early

West would be difficult to say. But can anyone hazard a guess that all the gold known to the Indians of Geronimo's day has been corraled into the known mines of Arizona?

The story of a modern Tantalus of the Apache retreats reminds us that the abundance of gold legends in this, or any other specified locality, can scarcely exist without considerable foundation. In mineral circles of Phoenix an interesting tale is still being related of a young white man held prisoner by Apaches, presumably in the Tonto territory northeast of that city. This region lies to the north of the Superstition Mountains, supposed site of the Lost Dutchman. It was before the Civil War when a young man was captured and kept with the tribe for an indefinite period. Possibly he was either a qualified "medicine man" or a reasonable facsimile thereof, for the Indians use for him. They even liberated him, scalp and all, in due time, and on that occasion took him, blindfolded, to a canyon, where he was treated to the sight of a vein of gold-bearing quartz, obviously rich and extensive, so the story goes. But he was not permitted to take any specimens of ore. Perhaps the Indians knew the tell-tale clues visible to a mining man in possession of a good sample.

In the light of history the blindfold appears to have been insult added to injury, in a country where capable prospectors with eyes open have lost good lodes. If this send-off to the parting guest shows a touch of Indian humor it is also true that the Apaches were taking no chances. Of course the victim searched long and diligently, after the end of the wars, but had to be content with having learned just how Tantalus felt, with the prize almost in reach.

The Lee Lost Lode

Somewhere in the silence of eerie canyons in the folds of a tortuous mountain range not too far from San Bernardino, California, there is a mine shaft that has not heard the tread of man since the decade of the 1890's. Possibly, and even probably, a skeleton is lying at the bottom of that shaft. The shaft, the labors of only two men, could not be of great extent, and its vein of gold-bearing quartz should be readily evident to the informed searcher.

Opinions vary as to the locality of this mine, lost for half a century, but many are inclined to credit persistent rumors that it may have been in the Bullion Mountains. This primitive range lies within south central San Bernardino County, along the north and east of Deadman Lake. Although within the bounds of a mineralized region, these mountains remain almost untouched by man, their sun-drenched crags still afford lookout for occasional mountain sheep, on guard for the tell-tale scent of man and oblivious of knowledge that game laws now protect them. Some remote cleft in the folds of these tumbled ridges rising from the waste of sand and weird Joshua Trees might conceal the little shaft. Who will learn the answer?

True to the traditions of lost gold in its tragic ending, the story of the Lee Lost Mine differs much in other particulars. Lee did not try to hide his mine. He even went so far as to record it at San Bernardino, describing its location as best he could—a description that was of no assistance to future searchers for quartz vein, shaft or camp. And Lee did a most unusual thing. He invited friends to come out and see the place. One of these was ex-Governor R. W. Waterman, not only a leading citizen but also an experienced mining man, promoter and organizer whose knowledge of ore qualified him to pass an opinion later, on the possibilities of the Lee diggings.

Lee, however, firmly refused to sell shares, and as the mine was at the end of a long and difficult journey in that unmotorized era, it is not surprising that there is no record of a visit by an outsider who might thus have found the place again. Only Lee's "hired man" knew the trails. The two men reduced the ore by means of a crude arrastra of their own make. Lee, or Old Man Lee, as he was called, brought it to San Bernardino and took supplies back to camp. In all respects the opposite of the reticent bad man type, the sociable Lee met with numerous friends and it is through their knowledge of his periodic shipments of gold ore and his out-spoken mention of his affairs that the existence of this mine is so well authenticated.

When fate overtook him, Lee had left San Bernardino one night for his digging. Travel at night afforded escape from the desert sun of the warmer season and did not indicate a need for secrecy. But Lee was found dead not far from town the next morning, shot through the heart. Robbery apparently was not the motive, for his clothing still contained money and his valuable watch. The murderer was not apprehended and no one was able to offer a valid reason for the mysterious killing.

Lee had mentioned that his mining camp was low on supplies, but this situation scarcely accounts for the second tragedy of the Lee Lost Mine. For the hired man was never seen again. A searching party headed by a county official set out at once, after Lee's body was discovered, but no trace of the mine could be found. Its dump and the arrastra would surely have been in evidence if the searchers had chanced upon the right gulch or canyon entrance. Even a faint trail toward the camp would have caught the eyes of desert men, but that attempt and many others were of no avail. It is difficult to believe that the helper calmly remained at the camp to starve, so the opinion prevailed that he had likely met with disabling injury or death in a rock fall or some other of the accidents familiar to the history of early small-scale mining. It is also possible that one of the sudden cloudbursts which frequently cause a flash flood might have swept upon him with its wall of water, even destroying the camp at the same time. His actual fate remains a mystery.

Ex-Governor Waterman published a standing offer of $40,000 for a half interest in the mine without other conditions, after Lee's claim to the location had expired. But even with this extraordinary inducement the frantic searchers finally had to give up the quest.

Somewhere, out there in the wilds, a small mine shaft and dump would probably be the chief remaining relics of the Lee operations, with perhaps a few tattered remnants of the camp or the wreck of a cabin. But the whole of it might be hidden away in a gulch or blended against a vast terrain—a speck on a mountainside.

Coachwhip Publications
CoachwhipBooks.com

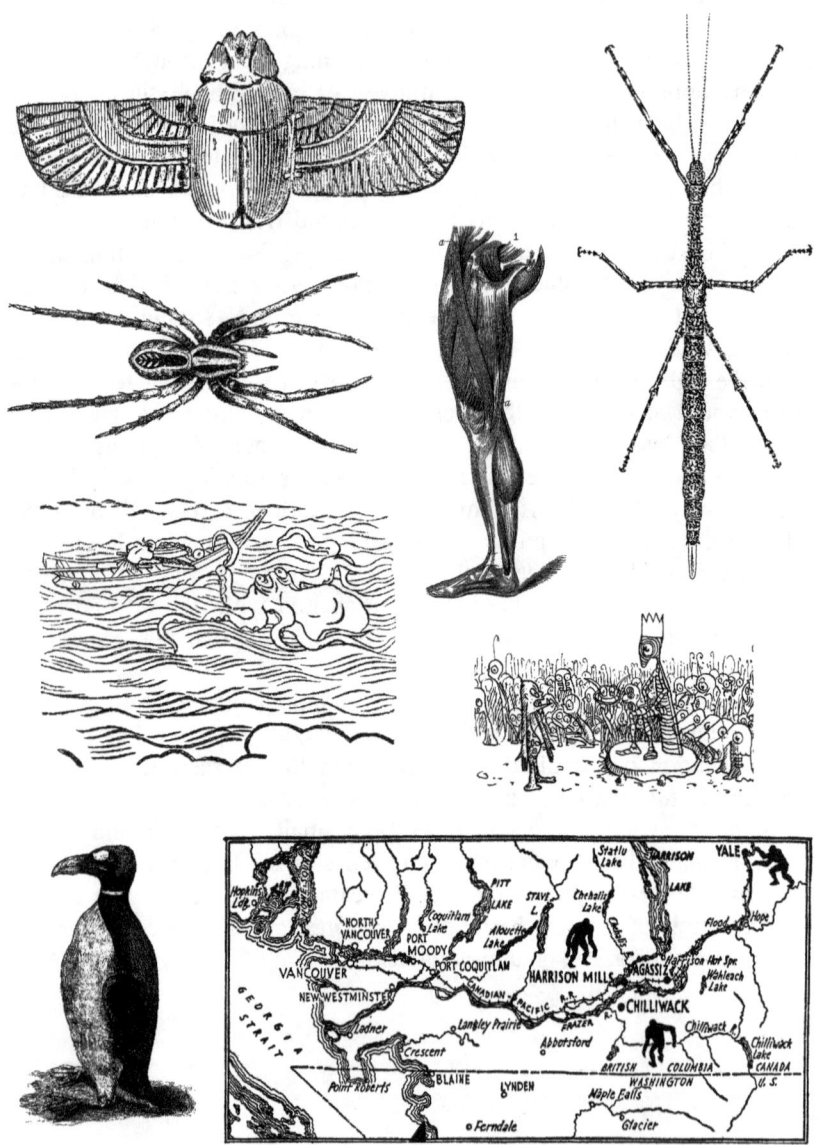

Coachwhip Publications

CoachwhipBooks.com

www.ingramcontent.com/pod-product-compliance
Lightning Source LLC
Chambersburg PA
CBHW061957070426
42450CB00011BA/3173